Chickens and Chicks

by Ann-Marie Kishel

first step nonfiction

Lerner Publications · Minneapolis

A chicken has a beak.

A chick has a beak.

A chicken has wings.

A chick has wings.

A chicken has feathers.

A chick has feathers.

Chickens and chicks are alike.